# NATURE WE NEED

# Why Do We Need POO?

by Laura K Murray

a Capstone company—publishers for children
www.raintree.co.uk

Raintree is an imprint of Capstone Global Library Limited, a company incorporated in England and Wales having its registered office at 264 Banbury Road, Oxford, OX2 7DY – Registered company number: 6695582

www.raintree.co.uk
myorders@raintree.co.uk

Hardback edition © Capstone Global Library Limited 2024
Paperback edition © Capstone Global Library Limited 2025
The moral rights of the proprietor have been asserted.

All rights reserved. No part of this publication may be reproduced in any form or by any means (including photocopying or storing it in any medium by electronic means and whether or not transiently or incidentally to some other use of this publication) without the written permission of the copyright owner, except in accordance with the provisions of the Copyright, Designs and Patents Act 1988 or under the terms of a licence issued by the Copyright Licensing Agency, 5th Floor, Shackleton House, 4 Battle Bridge Lane, London SE1 2HX (www.cla.co.uk). Applications for the copyright owner's written permission should be addressed to the publisher.

Edited by Donald Lemke
Designed by Sarah Bennett
Original illustrations © Capstone Global Library Limited 2024
Picture research by Svetlana Zhurkin
Production by Katy LaVigne
Originated by Capstone Global Library Ltd

978 1 3982 5384 1 (hardback)
978 1 3982 5385 8 (paperback)

**British Library Cataloguing in Publication Data**
A full catalogue record for this book is available from the British Library.

**Acknowledgements**
We would like to thank the following for permission to reproduce photographs: Dreamstime: Clara Cabezas, 20; Getty Images: alexei_tm, 25, Cloebudgie, 29 (bottom), fdevalera, 17, Gallo Images, 13, GregorBister, 16, izanbar, 9, JossK, 10, LA Waterhouse Photography, 21, mscornelius, 4, Whiteway, 29 (middle); Shutterstock: Alen Thien, 11, Alexey Morozov, 19, Clara Bastian, cover, David Calvert, 18, Faith Forrest (dotted background), cover and throughout, Francisco Martinez Lanzas, 12, hamdi bendali, 27, JamesChen, 14, Jeffrey B. Banke, 29 (top), jennyt, 24, kram-9, 28, Lens Quest, 7, Lorraine Logan, 5, Marco Tomasini, 8, Photo Fun, 6, Pixelheld, 8 (inset), pumab, 23, Reinhold Leitner (brown texture), cover, back cover, and throughout, Vladimir Wrangel, 15

Every effort has been made to contact copyright holders of material reproduced in this book. Any omissions will be rectified in subsequent printings if notice is given to the publisher.

All the internet addresses (URLs) given in this book were valid at the time of going to press. However, due to the dynamic nature of the internet, some addresses may have changed, or sites may have changed or ceased to exist since publication. While the author and publisher regret any inconvenience this may cause readers, no responsibility for any such changes can be accepted by either the author or the publisher.

Printed and bound in India.

# Contents

Leaving clues ................................................. 4

All about poo ................................................. 6

How does poo help us? ........................... 14

Threats to poo ............................................ 22

A world without poo ................................ 26

Cool facts about poo ............................... 29

      Glossary ................................................. 30

      Find out more .................................... 31

      Index ...................................................... 32

      About the author ............................. 32

Words in **bold** are in the glossary.

# Leaving clues

A black bear and her cub are close by. They left tracks in the forest. They also left piles of fresh poo. The poo contains seeds and bones. The bears have been eating berries and fish. Bear poo spreads seeds across the forest. The seeds will grow into new plants.

Everything that eats also poos. Poo has important jobs in **nature**. It helps plants grow. It gives humans clues about wildlife. Animals use poo in all sorts of ways.

## All about poo

Poo is solid waste. It's what is left of food that animals did not **digest**. Animal poo can contain hair, bones, scales, berries, seeds and bug parts.

earthworm

earthworm castings

An animal's body breaks down food. This helps the animal get energy and **nutrients** from the food. The rest passes out of its body.

Poo has many names. It may be called scat, droppings or dung. Bat poo is called guano. Caterpillar poo is known as frass. Earthworm droppings are castings.

Animal poo can be hard or soft. It can be smooth or lumpy. Poo has different smells too.

Some animal poo is shaped like a tube. Deer and rabbits poo small balls called pellets. Wombats poo in cubes!

wombat

whale poo

Poo can be brown, black, green, yellow or other colours. An animal's diet changes the colour of its poo. Whales eat lots of shrimp-like animals. This diet makes their poo pink, orange or red.

Animals use their poo in many ways. It helps them send messages to other animals. Some bury their poo. Others leave it in place. They use it as a warning to stay away.

Hippos mark their area with poo. They spin their tail. They spray clouds of poo underwater.

bird-dropping spider

Poo can keep animals safe. The bird-dropping spider looks like white poo from a bird. Other animals do not want to eat these spiders.

Some animals feed on poo. Egyptian vultures eat yellow cow dung. The dung has nutrients. It makes the birds' beaks bright yellow. A brightly coloured beak helps them find a mate.

Dung beetles feed on poo from cows, elephants and other animals. They roll dung into a ball and bury it. A dung beetle can bury dung that is 250 times heavier than itself!

Female dung beetles lay their eggs inside the balls. The poo is food for the babies. It keeps them safe too.

## How does poo help us?

Animals spread nutrients and seeds in their poo. Birds, bats and other animals eat fruit from plants and trees. The seeds come out in their droppings. They grow into new plants and trees.

tambaqui fish

Fish poo helps rivers, lakes and oceans. Tambaqui fish eat fruit and plant seeds. Their poo moves through the water. Then the seeds can grow into new plants in other places.

Humans study poo to learn more about animals. Scat gives clues about what animals are in the area. Poo can show if an animal is healthy or sick.

dinosaur dung fossil

Scientists study poo from millions of years ago. The poo helps them learn lots of information. People look at the size and shape of poo. They study poo through a microscope. The poo can show what food the animal ate.

Farmers and gardeners use animal poo as **fertilizer**. It is called manure. Manure adds nutrients to the soil. It can help plants and crops grow.

People may **compost** manure for a time. This breaks down the manure. It helps kill germs. It doesn't smell as bad either. Sometimes people make manure into dry pellets or they grow worms in the manure. Worm poo is rich in nutrients.

Humans use animal poo in many ways. People use livestock manure for fuel. The manure is made into gases and liquids. It gets turned into heat and electricity. People use **fibres** from animal poo to make paper and building materials.

Some people use dried poo to make fires.

llama poo

Poo can even be used to clean. Scientists use llama poo to clean dirty water. Helpful **bacteria** in the poo take out pollutants in the water.

## Threats to poo

All parts of nature are linked. They need each other. Today plants and animals have many threats.

The number of animals is falling. People are taking over the places animals live. **Climate change** is another threat to wildlife. There are fewer animals spreading seeds. It makes it harder for plants to grow.

Humans must be careful around animal poo. It carries germs. It can cause disease.

Raccoons and other animals may carry tiny roundworms. The eggs come out in the animals' poo. One dropping can have more than 10 million eggs!

Poo from cats, dogs and other pets can cause sickness too. People need to clean up after their pets. They should not feed or touch wild animals. They should wash their hands after being outside.

# A world without poo

Can you imagine a world without poo? Animals would not be able to get rid of their waste. They would not survive. The soil would be less healthy. Many plants and trees would not be able to grow. They would disappear. Food would be harder to find.

Animals would have to change how they live. They could not use poo for warnings or other messages. Dung beetles would not be able to eat or raise their young.

Every animal is different. So is their poo! Poo is an important part of nature. Humans, plants and animals need poo.

# COOL FACTS ABOUT POO

- Female wild turkeys poo in a spiral shape. Males poo in the shape of the letter "J".

- Some snakes may not poo for more than 400 days.

- Rabbits poo about 500 pellets a day. This would be like a human pooing once every three minutes.

- Vultures poo on their own legs to stay cool and kill germs.

- Sloths climb down from their tree about once a week to poo. They wiggle or dance before climbing back up.

- Otter dung (called spraint) can smell like violets.

# Glossary

**bacteria** very small living things that exist everywhere in nature

**climate change** significant change in Earth's climate over a period of time

**compost** mixture of decaying leaves, vegetables and other items that make the soil better for farming and gardening

**digest** break down food so it can be used by the body

**fertilizer** substance used to make crops grow better

**fibres** threads found in vegetables and other materials

**nature** world around us, including all of the plants, animals and other living things

**nutrient** substance needed by a living thing to stay healthy

# Find out more

## Books

*Animals (DKfindout!)*, DK (DK Children, 2016)

*Life Cycles: Everything from Start to Finish*, DK (DK Children, 2020)

*Under Your Feet (Underground and All Around)*, RHS and DK (DK Children, 2020)

## Website

**www.dkfindout.com/uk/animals-and-nature**
Find out more about animals and nature.

# Index

bacteria  21
bears  4
bird-dropping spider  11

climate change  22
compost  19

disease  24
dung beetles  13, 26

farming  18–19
fertilizer  18
fish  15
fuel  20

germs  19, 24

hippos  10
humans  16, 20, 24

livestock  20

manure  18–19, 20

names for poo  7
nutrients  7, 12, 14, 18, 19

safety  24
scientists  17, 21
seeds  4, 6, 14, 15, 22
soil  18, 26

vultures  12

# About the author

Laura K Murray is author of more than 100 published or forthcoming books for young readers. She loves learning from fellow readers and helping others find their reading superpowers!